MW01533742

Copyright@Henry Hollow2025

All rights reserved.
No portion of this book may be reproduced or utilized in any
form or by any electronic, mechanical, or other means without
the prior written permission of the publisher.

Written by: Henry Hollow
Ilustrated by Cita Aprilia K

First Edition Published in 2025
Printed in China
ISBN 978-1-962986-90-8
Publisher: Global Culture Press

NO HITTING, JUST HELPING!

What To Do When Big Feelings Come

What can my hands do? Hands can draw, clap, build, wave, and hug. They help me share and show I care. But hands are not for hitting—no, not there!

I used to hit when I was mad,
And that made others feel so bad.

But now I use my calm-down tools—one, two, three...eight! To help my big feelings fly away.

NO.1: When anger feels big and loud,
I take a deep breath and close my eyes.

NO.2: I hug myself so tight,
It feels like a blanket—soft and right.

NO.3: I stomp my feet
on the ground — not on someone!

NO.4: Try saying how I feel: "I am MAD!" and let them know.

NO.5: I squeeze my hands,
then open them slowly.

NO.6: I find my cozy, quiet place,
With my stuffed animal and a calm, safe space.

It's okay to feel mad. But hitting is not okay.
I can do so many good things with my hands!

Helping

Taking care

Sharing

My hands are not for hitting, not for hurting—no way!
They're for kind and gentle play.
When big feelings come, I know what to do!